Peripher

A Beginner's 3-Week Step-by-Step Plan to Managing the Condition Through Diet, With Sample Recipes and a 7-Day Meal Plan

Peripheral Neuropathy Diet

Copyright © 2022 Patrick Marshwell

All rights reserved No part of this book may be reproduced, or stored in a retrieval system, or transmitted in any form or by any means, electronic, mechanical, photocopying, recording, or otherwise, without express written permission of the publisher.

Printed in the United States of America

Disclaimer

By reading this disclaimer, you are accepting the terms of the disclaimer in full. If you disagree with this disclaimer, please do not read the guide.

All of the content within this guide is provided for informational and educational purposes only, and should not be accepted as independent medical or other professional advice. The author is not a doctor, physician, nurse, mental health provider, or registered nutritionist/dietician. Therefore, using and reading this guide does not establish any form of a physician-patient relationship.

Always consult with a physician or another qualified health provider with any issues or questions you might have regarding any sort of medical condition. Do not ever disregard any qualified professional medical advice or delay seeking that advice because of anything you have read in this guide. The information in this guide is not intended to be any sort of medical advice and should not be used in lieu of any medical advice by a licensed and qualified medical professional.

The information in this guide has been compiled from a variety of known sources. However, the author cannot attest to or guarantee the accuracy of each source and thus should not be held liable for any errors or omissions.

You acknowledge that the publisher of this guide will not be held liable for any loss or damage of any kind incurred as a result of this guide or the reliance on any information provided within this guide. You acknowledge and agree that you assume all risk and responsibility for any action you undertake in response to the information in this guide.

Using this guide does not guarantee any particular result (e.g., weight loss or a cure). By reading this guide, you acknowledge that there are no guarantees to any specific outcome or results you can expect.

All product names, diet plans, or names used in this guide are for identification purposes only and are the property of their respective owners. The use of these names does not imply endorsement. All other trademarks cited herein are the property of their respective owners.

Where applicable, this guide is not intended to be a substitute for the original work of this diet plan and is, at most, a supplement to the original work for this diet plan and never a direct substitute. This guide is a personal expression of the facts of that diet plan.

Where applicable, persons shown in the cover images are stock photography models and the publisher has obtained the rights to use the images through license agreements with third-party stock image companies.

Table of Contents

Disclaimer	**3**
Table of Contents	**5**
Introduction	**5**
Chapter 1: Peripheral Neuropathy: What Is It About?	**6**
Chapter 2: Prognosis and Treatment for Peripheral Neuropathy	**13**
Chapter 3: Managing and Prevention Through Diet	**16**
Chapter 4: Managing Peripheral Neuropathy Through Other Natural Methods	**25**
Chapter 5: A 3-Week Step-by-Step Plan	**31**
Sample Recipes	**38**
Chicken in Bouillon	38
Seared Salmon	42
Lemon Drop Cannellini Bean Soup	44
Grilled Eggplant	46
Chickpea Mint Tabbouleh	47
Sauteed Squash with Tomatoes and Onions	49
Kale Salad with Strawberry & Almonds	51
Avocado Chicken Lemon Salad	52
Lentil Stew	55
Vegan Spinach Soup	57
Barley Oat Pancakes	60
Salmon with Sweet Potato and Kale	62
Garden Vegetable and Herb Soup	65
Salmon Cakes	67

Conclusion   **69**
References and Helpful Links   **71**

Introduction

Peripheral neuropathy is a condition in which the peripheral nerves are damaged. These nerves are responsible for sending signals from the body to the brain. When they are damaged, they don't work properly and can cause a range of symptoms, including pain, numbness, and weakness.

The exact cause of peripheral neuropathy is often unknown, but it can be the result of diabetes, an infection, or another underlying health condition. There is no cure for peripheral neuropathy, but there are ways to manage the symptoms. One way is through diet.

A peripheral neuropathy diet can help to control symptoms and may even help to reverse nerve damage. This diet is high in whole grains, fresh fruits and vegetables, lean protein, and healthy fats. It's low in refined carbs, sugar, and saturated fat.

In this beginner's quick start guide, you will discover...

- All about peripheral neuropathy
- Its symptoms, causes, treatments
- Different ways to manage the condition
- Diet-friendly program for peripheral neuropathy

- A meal plan and a step-by-step plan to manage peripheral neuropathy

Chapter 1: Peripheral Neuropathy: What Is It About?

Peripheral neuropathy is damage to the nerves that carry information from the brain and spinal cord to the rest of the body. The word peripheral refers to the fact that these nerves lie outside of the brain and spinal cord. Neuropathy is a general term used to refer to any kind of damage or disease of the nerves. The term *pathy* comes from a Greek word meaning "suffering" or "disease."

The brain and spinal cord make up the central nervous system. The central nervous system is responsible for sending information to the rest of the body. The peripheral nervous system consists of the nerves that carry this information from the central nervous system to the rest of the body.

The spinal cord is located in the center of the back and is protected by the bones of the spine. The spinal cord is a long, thin bundle of nerves that carries information between the brain and the rest of the body.

The peripheral nervous system is made up of many different kinds of nerves. These nerves can be divided into two main types: sensory nerves and motor nerves

The peripheral nervous system is made up of three different types of nerves:

1. Sensory nerves carry information about touch, temperature, pain, and other sensations from the body to the brain.

2. Motor nerves carry information from the brain to the muscles, telling them when to contract.

3. Autonomic nerves carry information between the brain and the organs, glands, and other systems that control automatic functions like blood pressure, heart rate, and digestion.

When one or more types of peripheral nerves are damaged, it is called peripheral neuropathy. This damage can cause problems with sensation, movement, and other automatic functions.

Symptoms

The most common symptom of peripheral neuropathy is pain. Pain may be sharp, burning, tingling, or numb. It can occur in any part of the body, but it is most often felt in the hands, feet, arms, and legs. The reason pain is the most common symptom is that pain signals travel from the peripheral nerves to the brain. Other symptoms may include the following:

Muscle weakness: The muscles may feel tired, heavy, or hard to move.

- Loss of sensation: You may not be able to feel pain, heat, cold, or touch. You may also have trouble with balance and coordination.

- Numbness: You may feel like you have "pins and needles" in your hands or feet. There might be a prickly or tingling sensation.

- Loss of reflexes: Your reflexes, such as the knee-jerk reflex, may be reduced or absent.

- Changes in automatic functions: You may have problems with blood pressure, digestion, urination, and sexual function.

- Sensations of wearing gloves or socks

- Loss of coordination: You may feel clumsy or unsteady.

- Trouble walking: You may stagger or lose your balance.

- If the damage to the nerves is severe, you may also have problems with muscle wasting (atrophy), paralysis, and organ dysfunction.

- Pain: As mentioned earlier, pain is the most common symptom of peripheral neuropathy. The pain can range from mild to severe and can be constant or intermittent. It is often described as a burning, tingling, or numb sensation.

Causes

There are many different causes of peripheral neuropathy. The most common cause is diabetes.

Diabetes is a chronic disease that occurs when the pancreas is no longer able to produce insulin, or when the body cannot effectively use the insulin it produces. Insulin is a hormone that regulates blood sugar. When blood sugar levels are too high, it can damage your organs and lead to serious health complications. The pancreas is a small organ located behind the stomach. It produces insulin, which is a hormone that helps the body to use sugar for energy.

Diabetes can damage the nerves by causing them to become inflamed and swollen. The inflammation can cause the myelin sheath (the fatty covering that protects nerve fibers) to break down. This process is called demyelination. When the myelin sheath is damaged, it interferes with the ability of the nerves to send signals.

There are two main types of diabetes: type 1 and type 2. Type 1 diabetes, which used to be known as juvenile diabetes or insulin-dependent diabetes, is usually diagnosed in children, teens, or young adults. In type 1 diabetes, the body does not produce insulin.

Type 2 diabetes, which used to be known as adult-onset diabetes or non-insulin-dependent diabetes, is the most common form of diabetes. It usually develops in middle-aged or older adults, but it can also occur in children, teens, and young adults. In type 2 diabetes, the body does not produce enough insulin or the body's cells do not use insulin properly.

Besides diabetes, other causes of peripheral neuropathy include the following:

Autoimmune diseases occur when your immune system attacks your own organs and tissues. Examples of autoimmune diseases include lupus, rheumatoid arthritis, and celiac disease. Autoimmune diseases can be difficult to diagnose because they often mimic other conditions. Treatment for autoimmune diseases typically involves medications to suppress the immune system. In some cases, surgery may be necessary to remove damaged tissue.

Viral and bacterial infections can damage the nerves. Examples of viral infections include shingles, HIV, and the Epstein-Barr virus. Examples of bacterial infections include Lyme disease and leprosy.

Physical trauma, such as a car accident or a fall, can damage the nerves.

Cancer can directly or indirectly damage the nerves. Cancer that presses on the nerves or invades the nervous system is an example of direct damage. Chemotherapy and radiation therapy can cause indirect damage by damaging the blood cells that support the nerves.

Tumors that grow on or near the nerves can damage them. Examples of tumors that can cause peripheral neuropathy include acoustic neuromas and schwannomas.

Nutritional deficiencies

Not getting enough vitamins and minerals can lead to peripheral neuropathy. Vitamin B12 deficiency is a common cause of nerve damage.

Exposure to toxins certain chemicals, such as pesticides, heavy metals, and industrial chemicals, can damage the nerves. Alcohol abuse is another example of exposure to a toxin that can cause nerve damage.

Inherited disorders

Some types of nerve damage are inherited. Examples of inherited disorders that can cause peripheral neuropathy include Charcot-Marie-Tooth disease and Fabry disease. Charcot-Marie-Tooth disease (CMT) is a group of inherited disorders that affect the peripheral nerves—the nerves outside of the brain and spinal cord. The term "Charcot-Marie-Tooth" disease was first used in 1886 by three physicians—Jean-Martin Charcot and Pierre Marie in

France, and Howard Henry Tooth in England—who independently described the same disorder.

Given that there are many potential causes, people most at risk are those with

- A family history of the condition
- Excessive sugar or alcohol consumption
- Exposure to toxins
- A sedentary lifestyle, which can lead to obesity and type 2 diabetes

Chapter 2: Prognosis and Treatment for Peripheral Neuropathy

Diagnosis

Peripheral neuropathy is diagnosed based on a medical history and a physical examination. A detailed history can help identify the underlying cause of the nerve damage. The physical examination can help determine if the nerve damage is caused by damage to the peripheral nerves or the central nervous system.

Tests that may be ordered include:

- Blood tests. Blood tests can help rule out autoimmune diseases, infections, and nutritional deficiencies as causes of nerve damage.

- Nerve conduction study and electromyography. These tests can help determine if the nerve damage is caused by damage to the peripheral nerves or the central nervous system.

- Skin biopsy. A skin biopsy can help rule out inherited disorders as a cause of nerve damage.

- Imaging tests. Imaging tests, such as MRI and CT scans, can help rule out tumors as a cause of nerve damage.

Treatment

The treatment for peripheral neuropathy depends on the underlying cause of the nerve damage.

- If the nerve damage is caused by an infection, the goal of treatment is to eliminate the infection. Antibiotics are used to treat bacterial infections, and antiviral drugs are used to treat viral infections.

- If the nerve damage is caused by an autoimmune disease, the goal of treatment is to reduce inflammation and control the immune system with medications. In some cases, surgery may be necessary to remove damaged tissue.

- If the nerve damage is caused by diabetes, the goal of treatment is to control blood sugar levels with medications.

- If the nerve damage is caused by exposure to toxins, the goal of treatment is to remove the toxin from the body and treat any resulting damage.

- If the nerve damage is caused by a tumor, the goal of treatment is to remove the tumor with surgery. If the tumor cannot be removed, radiation therapy or chemotherapy may be used to shrink the tumor.

Prevention

There is no sure way to prevent peripheral neuropathy, but there are some things you can do that may help reduce your risk. These include maintaining a healthy weight, eating a healthy diet, exercising regularly, and avoiding excessive alcohol consumption.

Chapter 3: Managing and Prevention Through Diet

As stated, diabetes is one of the most common causes of peripheral neuropathy. Therefore, managing diabetes is a key part of preventing and managing peripheral neuropathy. A variety of lifestyle changes can help manage diabetes, and a healthy diet is an important part of that.

Foods to Avoid

Certain foods can make managing diabetes more difficult and can contribute to the development of peripheral neuropathy. These foods include:

Alcohol can interfere with blood sugar levels and increase the risk of nerve damage. Examples include beer, wine, and hard liquor.

Processed foods are high in sugar and unhealthy fats, which can contribute to weight gain and type 2 diabetes. Processed foods often contain additives and preservatives, which can be harmful to your health.

Some examples include:

- Cookies, chips, crackers, and processed meats
- Canned fruits and vegetables
- Frozen foods
- Foods with high levels of sodium

- Sugar-sweetened beverages
- Refined grains
- White flour
- Processed oils

It is best to avoid processed foods as much as possible and eat fresh, whole foods instead. If you do consume processed foods, be sure to read the labels carefully and choose products with minimal ingredients.

Refined carbs are high in sugar and can cause blood sugar spikes. The reason why refined carbs are so bad for your health is that they're stripped of all their nutrients. This means that when you eat refined carbs, your body has to work harder to process them, which can lead to weight gain and insulin resistance.

Examples of refined carbs include:

- White bread
- Pasta
- Rice
- Cookies
- Cake
- Chips
- Soda

- Fruit juices

Saturated fats can contribute to weight gain and type 2 diabetes. Saturated fats are a type of fat that is found in many animal-based products, such as meat and dairy. They can also be found in some plant-based oils, such as coconut and palm oil.

Saturated fats can contribute to weight gain and type 2 diabetes. They do this by increasing the amount of fat in the blood and by making the body resistant to insulin.

Examples of saturated fats include:

- Red meat
- Butter
- Cheese
- Ice cream
- Whole milk
- Processed meats
- Coconut oil
- Palm oil

Trans fats are unhealthy fats that can increase the risk of type 2 diabetes. Trans fats are a type of unsaturated fat that has been chemically altered to make it more solid. This process is called "hydrogenation." This process is highly unhealthy and can lead to a variety of health problems, including diabetes.

Foods high in trans fat include:

- Fast food, fried food, and processed food
- Margarine
- Vegetable shortening
- Packaged baked goods
- Cookies
- Chips
- Crackers
- Frozen pizza
- Microwave popcorn

Mercury is a toxic metal that can damage the nervous system. Fish is the main source of mercury exposure. The type of fish that tends to be highest in mercury are large, predatory fish such as shark, swordfish, and tuna. Smaller fish such as sardines, herring, and anchovies tend to be lower in mercury.

Some examples of fish high in mercury include:

- Shark

- Swordfish
- Tuna (fresh and canned)
- Mackerel
- Tilefish
- Orange roughy
- King mackerel
- Marlin

While these fish can be a part of a healthy diet, it is important to limit your intake to no more than 3-6 ounces per week to avoid mercury exposure.

Foods to Eat

Now that you've learned about which foods to avoid, here are some foods that can help you manage your peripheral neuropathy.

Whole grains are a good source of fiber and nutrients. They can help stabilize blood sugar levels and reduce the risk of type 2 diabetes.

Examples of whole grains include:

- Oats
- Barley
- Buckwheat

- Quinoa
- Brown rice
- Wild rice
- Whole-grain bread

<u>Fruits and vegetables</u> are good sources of vitamins, minerals, and antioxidants. They can help protect against damage to the nerves and reduce inflammation. Eat a variety of fruits and vegetables every day to get the most benefit.

Some examples include:

- Leafy greens, such as kale and spinach
- Cruciferous vegetables, such as broccoli and Brussels sprouts
- Berries, such as blueberries and raspberries
- Citrus fruits, such as oranges and grapefruit
- Sweet potatoes
- Beans and legumes

<u>Omega-3 fatty acids</u> are a type of unsaturated fat that can help reduce inflammation. They are found in fatty fish, such as salmon, and in flaxseed, chia seeds, and walnuts. Aim to consume two servings of fatty fish per week or take a daily omega-3 supplement.

Remember to buy wild-caught or sustainably sourced fish to avoid mercury exposure.

Some examples of foods high in omega-3 fatty acids include:

- Salmon
- Sardines
- Herring
- Anchovies
- Flaxseed
- Chia seeds
- Walnuts

Magnesium is a mineral that is involved in over 300 biochemical reactions in the body. It can help reduce inflammation and pain. Good sources of magnesium include dark leafy greens, nuts, seeds, and whole grains. You can also take a magnesium supplement.

Some examples of foods high in magnesium include:

- Black beans
- Pumpkin seeds
- Pinto beans
- Chickpeas

Vitamin B12 is a nutrient that is essential for nerve function. A lack of vitamin B12 can lead to peripheral neuropathy, which is a condition that causes nerve damage. The reason why vitamin B12 is so important for nerve function is because it helps to produce myelin, which is a substance that protects nerves.

Myelin is produced by oligodendrocytes, which are cells that are found in the central nervous system.

When there is a lack of vitamin B12, the body is unable to produce enough myelin. This can lead to nerve damage and dysfunction. Vitamin B12 deficiency is most common in older adults because they are more likely to have a condition that affects their ability to absorb nutrients.

It can be found in animal products, such as meat, poultry, fish, and eggs. You can also take a vitamin B12 supplement.

Some examples of foods high in vitamin B12 include:

- Salmon
- Chicken
- Beef*
- Eggs

*Beef is actually a type of red meat, which has high saturated fat. While it can be a good source of B12, moderation in consuming it is highly recommended.

Chapter 4: Managing Peripheral Neuropathy Through Other Natural Methods

Beyond diet, other natural methods can help prevent and manage peripheral neuropathy. These include:

Exercise

Exercise can help improve blood circulation, reduce inflammation, and relieve pain. Here are some examples:

- Aerobic Exercise

This type of exercise gets your heart rate up and is often recommended for people with diabetes. It can help improve blood sugar control and reduce the risk of heart disease.

Examples of aerobic exercise include walking, jogging, swimming, and biking. The best way to get started is to find an activity that you enjoy and start slowly.

Here is a plan to help you get started with aerobic exercise:

Choose an activity that you enjoy. If you don't like running, don't run! Many other activities can provide the same benefits.

Start slowly. It is important to warm up your muscles and gradually increase your heart rate.

Set a goal. Once you are comfortable with your starting activity, set a goal for how often you will do it. It is important to be consistent with your aerobic exercise routine.

Stick with it! Like any new habit, it takes time and effort to make aerobic exercise a part of your life. But the payoff is worth it! Regular aerobic exercise can help you control your blood sugar, lose weight, and reduce your risk of heart disease.

- Strength Training

Strength-training exercises can help improve muscle mass and blood sugar control. They can also reduce the risk of heart disease and improve mental health.

Examples of strength-training exercises include lifting weights and using resistance bands.

- Yoga

Yoga is a type of exercise that involves stretching and deep breathing. It can help improve flexibility, balance, and circulation. Yoga can also reduce stress and anxiety.

Yoga is an ancient practice that can help you achieve physical, mental, and emotional balance. If you're a beginner, it's important to find a class and instructor that are right for you. Here are some tips on how to get started with yoga:

1. Choose a class that is appropriate for your level. If you're new to yoga, look for a beginners' class or an all-levels class that will allow you to learn the basics at your own pace. One way to find a class that's right for you is to ask a friend or family member who practices yoga to recommend a class or studio.

2. Listen to your body and go at your own pace. Yoga is not a competition. It's about listening to your body and doing what feels comfortable and safe for you.

3. Don't be afraid to ask questions. If you're unsure about something, ask your instructor for clarification.

4. Stay hydrated and take breaks when you need to. Make sure to drink plenty of water before and after your yoga practice, and take breaks as needed.

5. Practice regularly. The more you practice yoga, the more benefits you will experience. Try to find a time that works for you and stick with it.

Yoga can be a great way to improve your overall health and well-being. By following these tips, you can get started on your yoga journey and begin reaping the rewards. Namaste!

Avoiding Alcohol and Tobacco

Alcohol and tobacco use can worsen the symptoms of peripheral neuropathy. If you drink alcohol, do so in moderation. This means no more than two drinks a day for men and one drink a day for women.

If you smoke, quit as soon as possible. There are many resources available to help you quit smoking, such as nicotine patches, gum, and counseling.

Here's a plan on how to quit smoking:

1. Choose a quit date and stick to it.

2. Talk to your doctor about quitting. He or she can prescribe medication to help control withdrawal symptoms.

3. Get rid of all tobacco products in your home, car, and workplace.

4. Avoid places where people smoke.

5. Stay busy and distracted. Exercise, read, or spend time with friends and family.

6. When you have a craving, wait 10 minutes and it will usually pass.

7. Reward yourself! Set goals and celebrate when you reach them.

Quitting smoking is one of the best things you can do for your health.

Sleep

Getting enough sleep is important for overall health as well as peripheral neuropathy. It is recommended that adults get 7-8 hours of sleep each night.

There are a few things you can do to improve your sleep:

1. Go to bed and wake up at the same time each day, including weekends.

2. Establish a regular

There are a few things you can do to help improve your sleep:

1. Establish a regular sleep schedule and stick to it as much as possible.

2. Avoid caffeine and alcohol before bedtime.

3. Keep a cool, comfortable environment in your bedroom.

4. Avoid watching television or working on the computer in bed.

5. Get up and move around during the day to help improve sleep at night.

6. Practice relaxation techniques such as deep breathing or meditation before bedtime.

Peripheral neuropathy can be a difficult condition to manage. However, by getting enough sleep and following these tips, you can help improve your symptoms.

Chapter 5: A 3-Week Step-by-Step Plan

Now that you've learned some basic information about peripheral neuropathy and how to manage it through diet, it's time to put this knowledge into action. This 3-week plan could be a potential way to manage this condition.

Week 1

Focus on preparing your mind and body for the diet program you're about to do. This may actually take time, so if you feel like starting in small steps earlier, then go ahead. Most people find it difficult to just switch up their diet, immediately changing to healthier options from what they're used to. Try to find ways to make it work for you.

Instead of immediately replacing all your meals at once, do it gradually.

Start by assigning a specific meal per day or specific days for a week when you're supposed to eat healthy meals until you get used to this diet plan. This will help you adjust well to the diet changes, as well as the new routine of being conscious in preparing your meals.

Each day or week, start to assign more meals or days until you're able to transition fully to eating healthier meals for an entire week.

Do the same thing with exercise, especially if you're used to living a sedentary lifestyle. Slowly incorporate these exercise routines into your schedule and gradually work on them until you're used to doing them regularly.

Keep in mind that the goal of this step-by-step plan is to help you start a journey and stay motivated to stick to it for a longer time.

Here are some other things you can also do during the first week, especially after you've adjusted well at the beginning of your journey:

1. Eliminate all processed foods, sugary drinks, and alcohol from your diet.

2. Eat three meals a day and include a mix of protein, healthy fats, and other foods listed in the "Foods to Eat" section.

3. Incorporate more whole foods into your diet such as fruits, vegetables, and lean proteins.

4. Drink plenty of water throughout the day and avoid caffeine.

5. Get at least 7-8 hours of sleep each night.

6. Exercise for 30 minutes every day. This can be a brisk walk, swim, or gentle yoga class.

7. Practice relaxation techniques such as deep breathing or meditation for 10-15 minutes each day.

Week 2

For this week, you may still experience some difficulty with the changes you're going through, but at least you have experienced how it is for the first seven days. You may also start to see the benefits of these changes, so use those benefits as your motivation to stick to the program.

For this, you can focus more on doing activities to keep your body active and healthy. Exercising greatly benefits almost everyone. Focus on strengthening the body to help improve blood circulation, relieve pain, and to lessen inflammation.

Follow and stick to programs that work best for you and that you find enjoyable. If you feel the need to stay motivated, try to find exercise buddies or a community where you can do work with other people who have the same goals as you.

1. Continue to eat three meals a day including a mix of protein, healthy fats, and complex carbohydrates.

2. Try new recipes and foods to keep your meals interesting.

3. Incorporate more whole foods into your diet such as fruits, vegetables, and lean proteins.

4. Drink plenty of water throughout the day and avoid caffeine.

5. Get at least 7-8 hours of sleep each night.

6. Exercise for 30-60 minutes every day. This can be a brisk walk, swim, or gentle yoga class.

7. Practice relaxation techniques such as deep breathing or meditation for 10-15 minutes each day.

Week 3

Hopefully for this week, you're a lot more adapted to the healthier lifestyle as compared to the first week. It's expected that you're eating healthier now and following a more active lifestyle. While it may be the last step of this plan, it should motivate you to keep this lifestyle, so you can continue reaping the benefits of this healthy change.

1. Continue to eat three meals a day including a mix of protein, healthy fats, and complex carbohydrates.

2. Try new recipes and foods to keep your meals interesting.

3. Incorporate more whole foods into your diet such as fruits, vegetables, and lean proteins.

4. Drink plenty of water throughout the day and avoid caffeine.

5. Get at least 7-8 hours of sleep each night.

6. Exercise for 30-60 minutes every day. This can be a brisk walk, swim, or gentle yoga class.

7. Practice relaxation techniques such as deep breathing or meditation for 10-15 minutes each day.

By following this 3-week plan, you will be on your way to managing your peripheral neuropathy and improving your overall health. Remember to listen to your body and make changes to the plan as needed. And most importantly, don't give up! With time and patience, you can improve your symptoms and live a healthy life.

We all know that consistency is key, so why not manage to make an effort to stick to the things that help you feel better and consistently do them? This 3-step plan may seem too simple, but actually, trying and implementing each step could actually greatly benefit you in the long run.

7-Day Meal Plan

A weekly meal plan is beneficial in helping you watch what you eat, and make sure you're meeting your daily nutrition needs. You can either follow or modify this meal plan according to your preference.

	Breakfast	**Lunch**	**Dinner**
Day 1	Grilled Eggplant	Salmon with Sweet Potato and Kale	Chicken in Bouillon
Day 2	Kale Salad with Strawberry & Almonds	Chickpea Mint Tabbouleh	Seared Salmon
Day 3	Barley Oat Pancakes	Lemon Drop Cannellini Bean Soup	Macrobiotic Bowl Medley
Day 4	Avocado Chicken Lemon Salad	Sauteed Squash with Tomatoes and Onions	Vegan Spinach Soup
Day 5	Garden Vegetable and Herb Soup	Salmon Cakes	Lentil Stew
Day 6	Barley Oat Pancakes	Grilled Eggplant	Salmon Cakes

| Day 7 | Chicken in Bouillon | Macrobiotic Bowl Medley | Salmon with Sweet Potato and Kale |

Sample Recipes

Chicken in Bouillon

Ingredients:

- 6 pcs. chicken legs
- 2 tbsp. canola oil
- 2 shallot stalks, chopped finely
- 2 tsp. fennel seeds
- 1 cup apple juice
- 2-1/2 cups chicken stock
- 2 cups wheat grains, rye, spelt, or polished barley, well rinsed
- 2-1/2 cups green asparagus, each cut into 4 pieces
- 2-1/2 cups spinach, chopped roughly
- salt
- pepper

For the gastrique:

- 4 cups balsamic vinegar
- 1 cup brown sugar

Instructions:

1. To prepare the gastrique, use a heavy-bottom saucepan to caramelize the sugar. Add the balsamic vinegar. Let the caramel dissolve.

2. When the mixture becomes syrupy in consistency, transfer it to a bottle. Keep the bottle in the refrigerator.

3. Sauté the chicken legs until they turn golden brown.

4. Put in the apple juice, fennel seeds, and shallots. Cook for around 2 minutes.

5. Put in the stock. Bring the mixture to a boil.

6. Bring down the heat and let the mixture simmer for 30 minutes.

7. Put them in another pan and let them simmer for about 12 minutes.

8. Add the wheat grains to the chicken.

9. Put in the asparagus.

10. Season the mixture with salt and pepper.

11. Add gastrique according to your taste. Let the mixture simmer for 8 to 10 minutes.

12. Add the spinach just before serving.

Seared Salmon

Ingredients:

- 1-1/2 tbsp. canola oil
- 4 pcs. salmon filets, each filet about 1-inch thick
- 1 tsp. kosher salt
- 1 tsp. ground black pepper, 1 teaspoon
- 2/3 cups shallots, thinly sliced, 2/3 cup
- 3 cups cherry tomatoes, 3 cups
- 2 tbsp. balsamic vinegar
- 1/2 cup basil leaves, torn

Instructions:

1. Preheat the oven to 500°F.
2. Use foil when lining a rimmed baking sheet, then set it aside.
3. Put a tablespoon of canola oil in a large heavy-bottomed pan placed over high heat.
4. Sprinkle evenly half of the pepper and salt over the fish filets.
5. Cook the filets in the pan for 4 minutes until the sides are golden brown.
6. Transfer the filets, with seared sides up, onto the prepared baking sheet.

7. Put it in the oven and cook the filet for about 4 minutes or until you get the degree of doneness that you prefer.

8. Return the skillet to the stove, and add the remaining canola oil.

9. Add the shallots and sauté for a couple of minutes. Season with the remaining salt and pepper.

10. Add the cherry tomatoes and 1/3 cup basil. Cook until the tomatoes are soft, for about 2 minutes.

11. Add the balsamic vinegar. Stir and cook for about a minute.

12. Transfer the filets to a serving dish and top with the balsamic vinegar-tomato mixture. Garnish with the remaining basil.

13. Serve and enjoy while hot.

Lemon Drop Cannellini Bean Soup

Ingredients:

- 2 cans cannellini beans
- 4 cups vegetable broth
- 2 tbsp. olive oil
- 1 yellow onion, chopped finely
- 2 large carrots, chopped finely
- 1 tsp. kosher salt
- 1 tsp. freshly cracked black pepper
- 4 garlic cloves, minced
- 1 can chickpeas
- 3 tsp. fresh rosemary, chopped finely
- 1 bunch spinach, roughly chopped
- 1 lemon (zest and juice)
- 1/3 cup nutritional yeast

Instructions:

1. In a blender, combine 1 cup broth and 1 cup cannellini beans. Blend until smooth then set aside.

2. In a large stockpot, heat oil over medium heat. Add carrots, and onions until softened.

3. Add remaining cannellini beans, rosemary, and chickpeas; stir to combine for one minute.

4. Boil 3 cups broth, bean mixture, and chopped spinach for 25 minutes or until thick.

5. Stir in the juice and zest of lemon, and sprinkle in the nutritional yeast.

6. Ladle hearty soup into 6 individual bowls; add black pepper and rosemary for garnish.

Grilled Eggplant

Ingredients:

- 2 small eggplants or 1 large eggplant, around 1-1/4 to 1-1/12 lb. in total, sliced into half-inch-thick rounds
- 2 tbsp. extra-virgin olive oil
- salt

Instructions:

1. Preheat the grill using the medium-high setting.
2. Toss eggplant slices and olive oil in a bowl.
3. Sprinkle it with salt to taste.
4. Toss ingredients again.
5. Place eggplant slices onto the grill.
6. Turn over to the other side after about 4 minutes, or until charred spots have appeared on the underside.
7. Continue grilling until eggplant slices have become tender.
8. When storing, place into an airtight container once it has cooled down, and then refrigerate. Grilled eggplant can last for up to 4 days in a chilled condition.

Chickpea Mint Tabbouleh

Ingredients:

- 1 can 15 oz. chickpeas or garbanzo beans, rinsed and drained
- 1 cup bulgur
- 1 cup fresh or frozen peas, thawed
- 1/2 cup minced fresh parsley
- 1/2 tsp. salt
- 1/4 cup minced fresh mint
- 1/4 cup olive oil
- 1/4 tsp. pepper
- 2 cups water
- 2 tbsp. soft, sun-dried tomatoes, julienned and not packed in oil
- 2 tbsp. lemon juice

Instructions:

1. Boil water and bulgur in a large saucepan.
2. Reduce heat. Allow to simmer with cover, for about 10 minutes.
3. Stir in peas. Cover.
4. Leave to cook until bulgur and peas are tender.
5. Transfer to a large bowl.

6. Add in the remaining ingredients. Stir well.
7. Serve warm, or refrigerate and serve cold.

Sauteed Squash with Tomatoes and Onions

Ingredients:

- 1 medium onion, finely chopped
- 1 tsp. salt
- 1/4 tsp. pepper
- 2 large tomatoes, finely chopped
- 2 tbsp. olive oil
- 4 medium zucchini, chopped

Instructions:

1. In a large skillet over medium-high heat, heat oil.
2. Cook onion and stir until tender.
3. Stir in zucchini, and cook for 3 minutes.
4. Stir in tomatoes, salt, and pepper.
5. Cook squash until tender, about 4-6 minutes longer. Stir occasionally.
6. Serve with a slotted spoon.

Kale Salad with Strawberry & Almonds

Ingredients:

- 1 bunch of kale
- 1/2 cup sliced strawberries
- 1/4 cup sliced almonds
- 1 lemon pulp juice
- 1/8 tsp. salt
- 1/8 tsp. black pepper
- 1 tbsp. agave
- 2 tbsp. of olive oil

Instructions:

1. Rip kale into small pieces and massage with hands until tender.
2. Put it in a bowl. Add almonds and strawberries.
3. To create a dressing, mix lemon juice with olive oil, salt, pepper, and agave, and then pour it over the salad.
4. Serve immediately.

Avocado Chicken Lemon Salad

Ingredients:

- 2 organic chicken breast, skinless
- curly kale, a bunch, ribs and stems removed
- 1 cup of cooked wheat berries
- 1 ripe avocado, sliced, drizzle it with lemon juice
- 1/2 cup pomegranate arils
- 1/2 cup pine nuts, toasted
- pink peppercorns
- pea shoots

For the rosemary oil marinade:

- 1/2 lemon, zest only
- 1 sprig of rosemary
- 2 tbsp. olive oil
- sea salt
- black pepper

For the lemon vinaigrette:

- 1 tsp. dijon mustard
- 1-2 cloves of garlic, minced

- 2 anchovy filets, minced
- 1 small lemon, juice only
- 2 tbsp. extra virgin olive oil
- 1/2 tsp. lemon zest
- sea salt
- black pepper

Instructions:

1. Prepare the chicken by washing and draining with a paper towel.
2. Slice through the chicken breasts for the marinade and cook well later.
3. Using a mortar, mix all ingredients for the rosemary oil marinade until you get aromatic oil.
4. Gently rub the chicken with the rosemary oil and marinate for at least 15 minutes at room temperature or up to 8 hours in the refrigerator. Occasionally turn over the bag during the day.
5. Preheat the oven up to 375°F.
6. Heat cast-iron skillet over medium-high heat.
7. Add in chicken breasts. Cook until both sides are brown.

8. Move the skillet to the oven and cook for about 7-10 minutes.

9. Using a whisk, combine all the lemon vinaigrette ingredients in the bowl.

10. Put the kale and lemon vinaigrette in a large mixing bowl. Use your hands to mix for about a minute or two. Adjust seasoning according to your preference.

11. Move kale on a serving plate, topped with avocado slices.

12. Slice the chicken and place it on top of the salad. Top with peppercorns, pomegranate arils, toasted pine nuts, and wheat berries.

13. For garnishing, add pea shoots.

14. Enjoy by serving either warm or chilled, with the grilled lemon on the side.

Lentil Stew

Ingredients:

- 4 cups savoy cabbage, chopped
- 1 tbsp. anchovy paste
- 3-1/2 tbsp. extra virgin olive oil
- 2 tsp. cumin
- 1 shallot, chopped finely
- 1 tsp. turmeric
- 1 leek, chopped finely
- 4 carrots, chopped finely
- 2 celery stalks, chopped finely
- pepper
- unrefined sea salt
- 2 cups organic vegetable broth
- 1 26-oz. can chopped tomatoes, drained
- 2 15-oz. cans lentils, rinsed and drained
- 1 tbsp. unfiltered and unpasteurized apple cider vinegar
- 1/2 cup parsley, chopped

- 4 organic pasture-raised eggs
- optional: egg topping

Instructions

1. Boil water in a large pot.
2. Add cabbage and let cook for about 10 minutes, or until soft. Drain after and set aside.
3. In the same pot, pour in oil and place over medium heat.
4. Add the cumin, anchovy paste, shallots, turmeric, celery, carrots, and leeks.
5. Saute for about 8 to 10 minutes, or until the vegetables are soft.
6. Season with salt and pepper, according to your taste.
7. Add the tomatoes and cabbage, and let cook for another 5 minutes.
8. Pour in the vegetable broth and cook for 10 minutes.
9. Add lentils and cook for another 5 minutes.
10. Remove stew from heat, and add parsley.
11. Serve while hot.

Vegan Spinach Soup

Ingredients:

- coconut oil
- garlic
- onion
- celery
- green onion
- thyme
- vegetable bouillon
- coconut milk
- potato
- spinach
- vegetable broth
- allspice
- Scotch Bonnet pepper
- Optional: dumpling

Instructions:

1. On medium heat, stir-fry onion in oil until soft, about a couple of minutes.

2. Stir in garlic, celery, green onions, and thyme. Cook until it becomes fragrant.
3. Put in spinach, and cook until wilted.
4. Add coconut milk, vegetable bouillon, vegetable broth, allspice, dumplings, potatoes, and Scotch Bonnet. Wait until it boils.
5. Lower the heat, and leave to simmer for around 25-30 minutes.
6. Optional: add dumplings into the stew as it simmers.
7. Serve while hot.

Barley Oat Pancakes

Ingredients:

- 1 cup barley flour
- 1 cup oat flour
- 1 tbsp. baking powder, sodium-free
- 1 tsp. salt
- 1-1/2 cup nonfat milk
- 2 pcs. large eggs
- 2 tbsp. canola oil
- 2 tbsp. honey
- 2 tsp. vanilla extract
- honey or maple syrup, for serving
- your choice of fresh fruit

Instructions:

1. In a large mixing bowl, whisk together the oat flour, barley flour, salt, and baking powder.
2. In a separate mixing bowl, whisk together the eggs, oil, non-fat milk, vanilla extract, and honey.

3. Transfer the wet ingredients to the large mixing bowl. Whisk them together to combine. Do not overmix the batter.

4. Place a large non-stick pan over low-medium heat.

5. Put about 3 tablespoons of batter into the pan. Wait for bubbles to appear on the top side of the pancake and the bottom to turn golden brown.

6. Flip the pancake to cook the other side.

7. Repeat until all the batter is cooked.

8. Top each pancake with your choice of fresh fruit.

9. Drizzle honey or maple syrup over the fruit.

10. Serve the pancakes immediately.

Salmon with Sweet Potato and Kale

Ingredients:

- 1 lb. salmon
- 1 head of kale
- 1 sweet potato
- 1 tbsp. olive oil
- salt
- pepper

Instructions:

1. Preheat the oven to 350°F.
2. Remove the stem from the kale and chop it into small pieces.
3. Peel the sweet potato and cut it into small cubes.
4. Combine the salmon, kale, sweet potato, and olive oil in a baking dish. Season with salt and pepper.
5. Bake for 20-25 minutes, or until the salmon is cooked through.

Macrobiotic Bowl Medley

Ingredients:

- 1/2 cup brown rice
- 3 cups chard, roughly chopped
- 1 cup squash, diced
- 1 cup broccoli florets
- 1 cup black beans, thoroughly rinsed and drained
- 1 oz. kombu
- 1/2 cup sauerkraut, chopped

Sauce:

- 2 tbsp. sesame tahini
- 2 tbsp. sodium tamari
- 1 clove garlic
- 1 tbsp. ginger
- 1 lime, juiced

Instructions:

1. Boil 1 cup of water.
2. Add rice and allow it to boil. Cover and reduce heat and simmer for 40 minutes.

3. Remove from heat and allow to sit covered for another 10 minutes, then fluff with a fork.

4. Place beans in a pot with kombu. Cover with water, and bring to a boil.

5. Reduce heat and simmer for 15-20 minutes. Drain and rinse after.

6. Place a steamer basket in a pot with water and bring to a boil.

7. Add broccoli, cover and steam for 4-5 minutes then remove, keeping water in the pot.

8. Add squash, cover and steam for 4-5 minutes then remove, keeping water in the pot.

9. Add chard, cover and steam for 3-4 minutes, then remove.

10. Mix all the ingredients of the sauce.

11. Serve everything on a plate and enjoy!

Garden Vegetable and Herb Soup

Ingredients:

- 1 can 14-1/2 oz. diced tomatoes in sauce
- 1 medium yellow summer squash, halved and sliced
- 1 medium zucchini, halved and sliced
- 1 lb. red potatoes, cubed
- 1 tsp. dried basil
- 1/2 tsp. paprika
- 1/2 tsp. salt
- 1/4 tsp. dill weed
- 1/4 tsp. pepper
- 1-1/2 cups vegetable broth
- 1-1/2 tsp. garlic powder
- 2 cups water
- 2 large carrots, sliced
- 2 medium onions, chopped
- 2 tbsp. olive oil

Instructions:

1. In a large saucepan, heat oil over medium heat.
2. Add onions and carrots.

3. Cook and stir until onions are tender.

4. Add potatoes and cook for 2 minutes.

5. Stir in water, tomatoes, broth, and seasonings. Let it boil.

6. Reduce heat. Remove cover and allow to simmer until potatoes and carrots are tender.

7. Add yellow squash and zucchini. Cook until vegetables are soft.

8. You can serve this as is.

9. Another option before serving, in batches, puree the mixture until you get the preferred consistency. Add broth if necessary.

Salmon Cakes

Ingredients:

- 1 lb. sockeye salmon, ground using a food processor
- 1/4 cup mayonnaise, vinegar-free
- 1 tsp. sea salt
- 1/2 pc. zucchini, chopped finely
- 1 tsp. chives or parsley, dried or fresh
- 1/2 cup rice crumbs, or grind rice cakes in a food processor
- 1 egg
- rice bran oil

Vinegar-free mayonnaise:

- 2 free-range eggs
- 1 tsp. wasabi mixture, dissolve wasabi in a little bit of warm water
- 1 tsp. yellow mustard powder
- 3/4 cup rice bran oil
- sea salt
- 2 tbsp. olive oil
- 1 lemon, juice only
- 1 tbsp. ume vinegar

- optional: 1 tbsp. soy lecithin or 1 egg yolk

Instructions:

1. Blend in the ground salmon, the mayo, zucchini, half of the rice crumbs, sea salt, and herbs, the last two added according to taste.
2. Make 3-inch patties with the mixture.
3. Beat the egg.
4. Dip the patties in the egg followed by the rice crumbs.
5. Fry the patties, for about 5 minutes on each side.
6. Serve while warm.

To make the mayonnaise:

1. Blend the first three ingredients in a blender, with 1/4 cup of rice bran oil and a pinch of sea salt.
2. While blending, gradually pour the remaining rice bran oil.
3. Add the remaining ingredients, except for the yolk or soy lecithin.
4. For a thicker texture, add the soy lecithin. For a runnier texture, add the yolk. Continue to blend until the preferred thickness is achieved.
5. Put the homemade mayo into a mason jar and consume it for two weeks.

Conclusion

Peripheral neuropathy can be a difficult condition to manage, but it's important to remember that you're not alone. There are many resources and treatments available to help you cope with this condition.

The most important thing you can do is to educate yourself about peripheral neuropathy and learn as much as you can about the condition. This will help you better understand your symptoms and find the treatments that work best for you.

In addition to learning about peripheral neuropathy, it's also important to eat a healthy diet and get plenty of exercise. These lifestyle changes can help improve your symptoms and overall health.

Finally, don't forget to reach out to your doctor or a support group if you need help. There are many people who understand what you're going through and can offer helpful advice and support.

If you found this guide helpful, please leave a review or rating. Best of luck on your journey to managing your peripheral neuropathy.

References and Helpful Links

"7 Foods That Help Fight Neuropathy." Niva Health, 8 Oct. 2021, https://www.nivahealth.com/2021/10/08/7-foods-that-help-fight-neuropathy/. Accessed 3 July 2022.

Neuropathy Nutrition 101 - Neuropathic Therapy Center | Loma Linda University Health. 2 May 2018, https://lluh.org/services/neuropathic-therapy-center/blog/neuropathy-nutrition-101. Accessed 3 July 2022.

Peripheral Neuropathy. 8 Aug. 2021, https://www.hopkinsmedicine.org/health/conditions-and-diseases/peripheral-neuropathy. Accessed 3 July 2022.

"The Foundation For Peripheral Neuropathy." The Foundation For Peripheral Neuropathy, https://www.foundationforpn.org. Accessed 3 July 2022.

The Link Between Poor Nutrition and Neuropathy: David Berkower, DO: Physical Medicine and Rehabilitation. https://www.painandspinerehab.com/blog/the-link-between-poor-nutrition-and-neuropathy. Accessed 3 July 2022.

Walker, Jonathan. "6 Types of Food People With Neuropathy Should Avoid." Ethos Health Group, https://ethoshealthgroup.com/food-people-with-neuropathy-should-avoid/. Accessed 3 July 2022.

Printed in the USA
CPSIA information can be obtained
at www.ICGtesting.com
LVHW010211230224
772636LV00008B/469